Deadly Creatures of the Sea

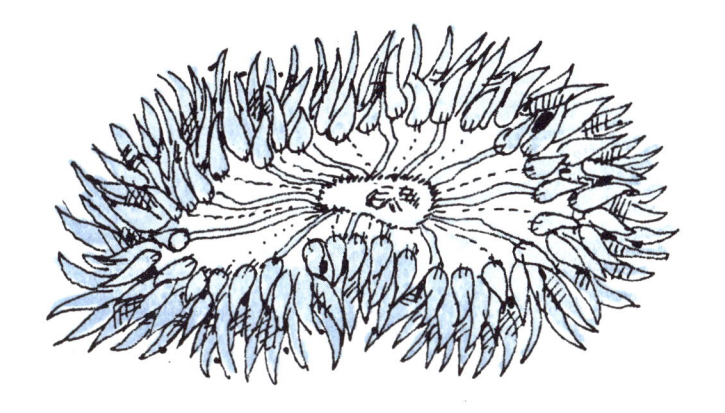

by WALTER HARTER

illustrated by ALBERT MICHINI

A Finding-Out Book
Parents' Magazine Press • New York

Library of Congress Cataloging in Publication Data
Harter, Walter L
   Deadly creatures of the sea.

   (A Finding-out book)
   Includes index.
   SUMMARY: Discusses nine deadly ocean creatures often found close to the shore.
   1. Dangerous marine animals–Juvenile literature.
[1. Dangerous marine animals. 2. Marine animals]
I. Michini, Albert.   II. Title.
QL100.H36   591.6'9'091634   76-26541
ISBN 0-8193-0841-2

Text copyright © 1977 by Walter Harter
Illustrations copyright © 1977 by Albert Michini
All rights reserved
Printed in the United States of America

Contents

INTRODUCTION 5

THE STARFISH 7

THE PORTUGUESE MAN-OF-WAR 13

THE SCORPION FISH 18

SEA ANEMONES 25

THE OCTOPUS 32

CORALS 38

THE TILEFISH 44

THE LAMPREY 50

SEA URCHINS 56

INDEX 63

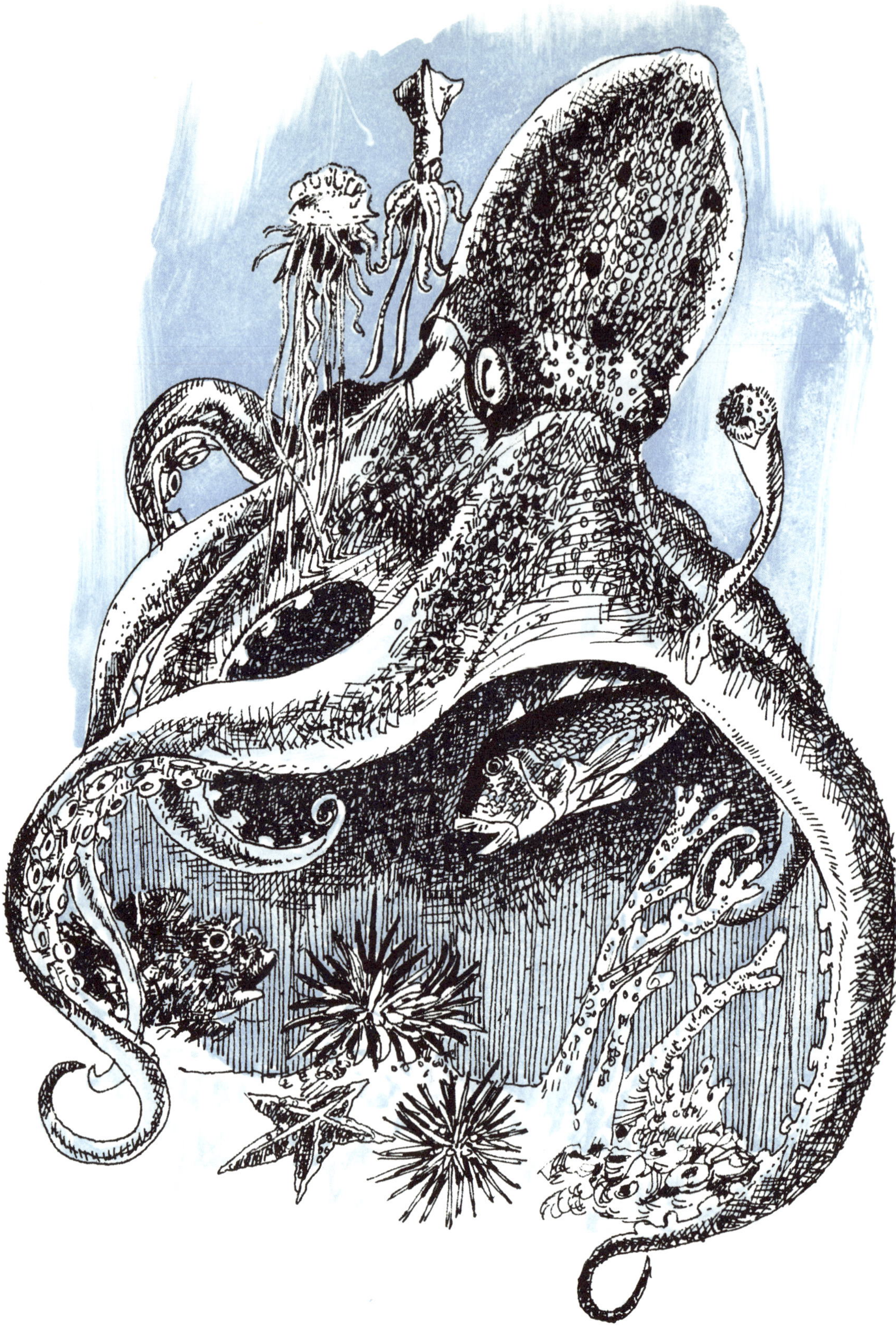

Introduction

Because the sea is so beautiful we sometimes forget that many dangerous creatures live beneath its surface. Some of them hide in secret places along the shores. Others live in the deep waters.

Most people know about the large, dangerous animals that live in the oceans. There are also small creatures which can be as harmful to people. But most of them are dangerous only when they are disturbed.

Many of these small sea animals use a poison to capture their food, or to defend themselves against enemies who want *them* for food.

Some of these creatures sting, others bite. Some of them stay in one place all their lives. Others move swiftly or slowly through the oceans of the world.

Starfish

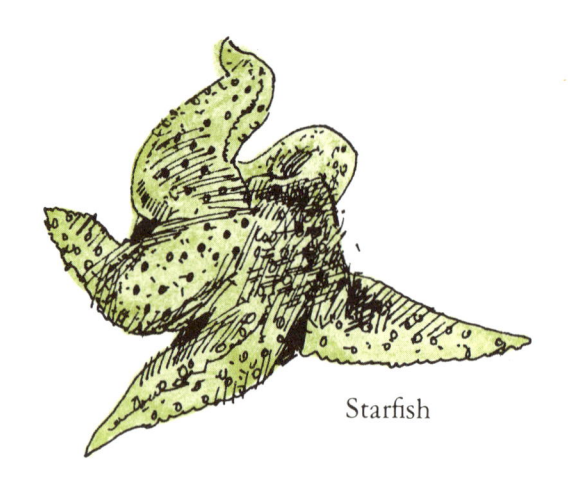

Harmful as these small sea animals can be, each one is an important link in the chain of life that holds the smallest and largest living things together.

Many of these creatures live only in faraway oceans. Some of them live in all the seas. In this book we will talk about a few of them you might meet along the seashores of North America.

Starfish

Oyster

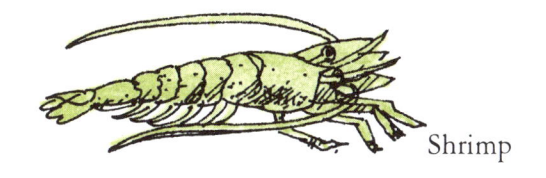

Shrimp

The Starfish

The starfish is called the garbage man of the ocean because it crawls over the sea floor and eats anything it touches. Fishermen call it an enemy because it will eat whole beds of oysters and shrimps.

Starfish find their food by stumbling over it. Although they have an eye at the tip of each of their five arms, they can't see very well.

When a starfish finds an oyster or a shrimp, or other shellfish, it wraps its arms around the shell and holds it. Then the tiny spikes on each arm shoot poison into the creature and stun it, while the five arms begin to pull the shell apart.

Starfish with clams

Starfish are also called "walking stomachs," because of the strange way they eat their food. They don't eat with their mouths, as all other creatures do. A starfish's mouth is only a tiny slit in the center of its body. When it has found something to eat, it pushes its entire stomach out through this small opening and surrounds the food with it. When the food has been digested, the stomach is pulled back through the small mouth.

Barnacles

Sometimes large numbers of starfish will roll themselves into a ball and tumble across the sea floor until they bump into a bed of oysters or shrimps. Then the ball will break apart and each starfish will begin to dine on shellfish.

Starfish aren't good food for humans. Many scientists who study sea creatures think that starfish aren't good for anything. But starfish do serve a purpose by eating the tiny sea animals, called barnacles, that cling to the bottoms of ships. They also eat the "oyster borer," a worm that kills many oysters. So perhaps the starfish saves the lives of many more shellfish than it eats.

A female starfish gives birth to thousands of children at one time. The babies are so small that ten of them could be placed on the head of a pin.

In three weeks' time the young starfish grow to more than fifty times the size they were when they were born. Some starfish are only as large as a quarter of a dollar when they are fully grown. Other starfish grow to be giants that measure a yard from the tip of one arm to the tip of an arm on the opposite side of their bodies.

One of the amazing things about a starfish is that it can grow new arms when one or more of them has been lost in fights with other creatures, or by accident. And if one of the lost arms has only a small piece of the body attached to it, the starfish can grow an entirely new and whole body, complete with five arms!

As long as a starfish remains in the ocean, it has few enemies. But often it is cast up onto the shore by waves and tides. Then seagulls and pelicans attack it. When a starfish is rolled over onto its back, it can't use its poisonous spikes, and therefore can't defend itself. Then the sea birds have a feast.

Starfish are of many colors—blue, green, yellow, and some that live in the deepest waters are blood-red. But when they are dead, and have dried, they become white.

Starfish of various kinds and sizes are found in all the oceans and seas. But the most common type is in the Atlantic Ocean from Maine to Florida, and in the Gulf of Mexico.

They move very slowly across the sea bottoms, traveling only three to six inches a minute, or from two to four miles a month.

Starfish can't live very long when they are out of the water. Those found lying on a beach are usually dead, but their spikes might still contain poison. If they are picked up, it should be done carefully, and only if gloves are worn.

The spikes of starfish are even more harmful when the creatures are alive. Be careful when swimming or wading in the shallow waters of the seashore. If a starfish is seen lying or crawling on the sandy bottom, it is dangerous to step on it or to touch it with your fingers.

The Portuguese Man-of-War

It looks like a small boat with a pretty blue sail as it bobs along on top of the waves. Hanging down into the water from its bottom are dozens of streamers, some as long as fifty feet.

This is a member of the large jellyfish family—the Portuguese man-of-war, one of the most interesting and dangerous creatures of the sea. The name comes from the fact that its sails look like those of the Portuguese and Spanish ships that sailed the seas many hundreds of years ago.

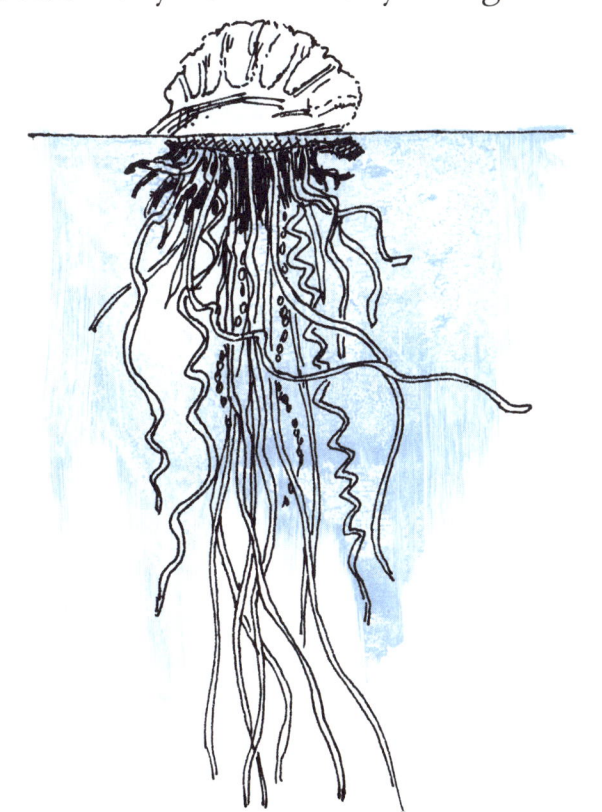

The sail of the Portuguese man-of-war is a small balloon filled with gas. When the creature sees food under the water, it puffs out the gas and sinks below the surface to eat. When it wants to move on, it refills its balloon-sail with the gas it makes in its stomach and pops up onto the surface again.

Although the Portuguese man-of-war appears to be one complete creature, it is really a combination of many parts. It is a colony of various animals that have learned to live together. However, they all come from the same egg.

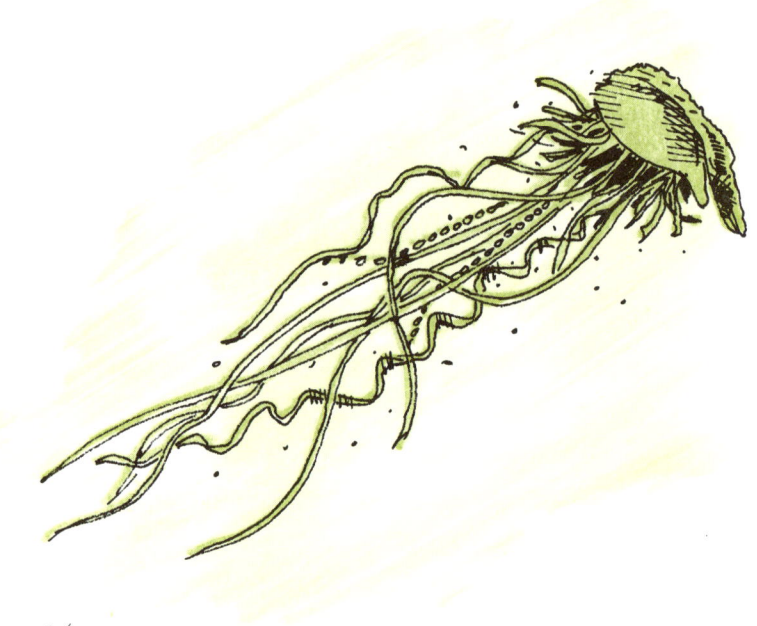

Nomeus

The streamers perform many tasks. Some of them are covered with tiny barbs, like arrows. These are shot into any fish, or person, that swims close by. The darts are very poisonous.

Different streamers that hang into the water carry the food, which has been killed by the barbs, to other streamers. These are covered with small, sticky mouths that quickly eat and digest the food. New barbs are grown again so that more fish can be caught.

One of the strange things about the Portuguese man-of-war is that some very small fish, called nomeus, keep it company wherever it drifts.

These nomeus swim in and out among the

streamers without being hurt. They seem to act as bait, teasing larger fish to chase them within reach of the dangerous arrows, or barbs.

Portuguese men-of-war are found in all the warm seas of the world. They appear to drift wherever the winds push their small sails. Many of them are found along the shores from New England to Florida.

The Portuguese man-of-war looks so pretty and harmless that swimmers often try to pick one up when it floats by them. But the creatures should be avoided whenever they are seen.

If one of the creature's weapon streamers wraps itself around the arm or leg of a swimmer, it causes great pain. If the swimmer is attacked by several streamers, that person should be taken to the hospital immediately. Sometimes swimmers have found themselves surrounded by a fleet of the little blue sails. If a swimmer is attacked by many of the Portuguese men-of-war the person usually dies before reaching shore.

The Portuguese men-of-war seldom travel alone. They are usually in groups of from 20 to 100. And they move so silently and swiftly through the water that they can appear in an area as if by magic. They do not deliberately attack humans, but will shoot their darts into anything that comes close to them.

Even when Portuguese men-of-war are cast up onto the beach, their streamers are still dangerous. But after they have lain on the sand for several hours, they seem to melt. Then what is left looks like pieces of discarded plastic sandwich bags.

Scorpion

The Scorpion Fish

The scorpion fish is found in many of the warm seas of the world. Some of them live in the Gulf of Mexico and along the west coast of Florida.

The fish didn't get its name because it looks like the scorpion that lives on land, but because the poison it carries in the spikes on its back is as harmful as the poison in the sting of the black bug that lives in tropical countries.

The scorpion fish is one of the prettiest of all the sea creatures. It is a mixture of many colors—red, white, blue, green, and yellow. These bright colors help the scorpion fish to hide among the beautifully tinted corals and not be seen by its enemies, or by the smaller fish it catches for food.

Scorpion fish

Coral garden

And that is the most dangerous thing about this pretty sea animal. When it burrows into the sandy bottoms of the shores, it seems to disappear into its surroundings. Only the waving fins and spikes along its back stick up from the sand.

Sometimes people who are wading in shallow water accidentally step on these spikes and cut themselves. The poison of the scorpion fish is so deadly that unless the wound is taken care of at once the person becomes very ill, and sometimes dies.

But those pretty fins and spikes help the scorpion fish catch its food. For, just as a fisherman will bait a hook to attract a fish, the scorpion fish has its own bait and lure to coax a fish within reach of its poison.

Not all the fins and spikes on the back of a scorpion fish are poisonous. One large fin, called the dorsal fin, has a special purpose. When a scorpion fish sees a small fish swimming nearby, it lies very still in the sand. Then the dorsal fin begins to rise higher than the others. As this middle fin grows larger it becomes bright red. At the same time a dark band appears at the bottom of the fin, making it seem to be separated from the back of the scorpion fish.

As the dorsal fin becomes even more red, a black spot appears on it that looks like an eye. Then another dark line looks like a mouth. In a few seconds the dorsal fin has all the appearance of a very attractive small red fish.

But this imitation fish has not yet completed its work as a lure. It begins to move back and forth, faster and faster, to attract the attention of the small fish. Slowly the small fish swims closer and closer. At exactly the right moment, one of the poisoned spikes strikes it. When the small fish has been stunned by the poison, the large mouth of the scorpion fish opens and snaps it up.

As soon as the small fish has been caught, the imitation fish on the back of the scorpion fish shrinks back and becomes the dorsal fin again. The bright red color fades away, and once again the scorpion fish lies quietly in the sand, waiting for another small fish to swim by.

Although the scorpion fish has been studied for many years, no one yet has been able to understand how the dorsal fin becomes an imitation fish so quickly, and then disappears when a fish is caught. It remains one of the mysteries of sea life.

Because it stays very close to the bottom of warm seas, the scorpion fish is dangerous to humans only when it is stepped on by a wader, or if it is touched by a diver who might swim too close to the corals in which it is hiding.

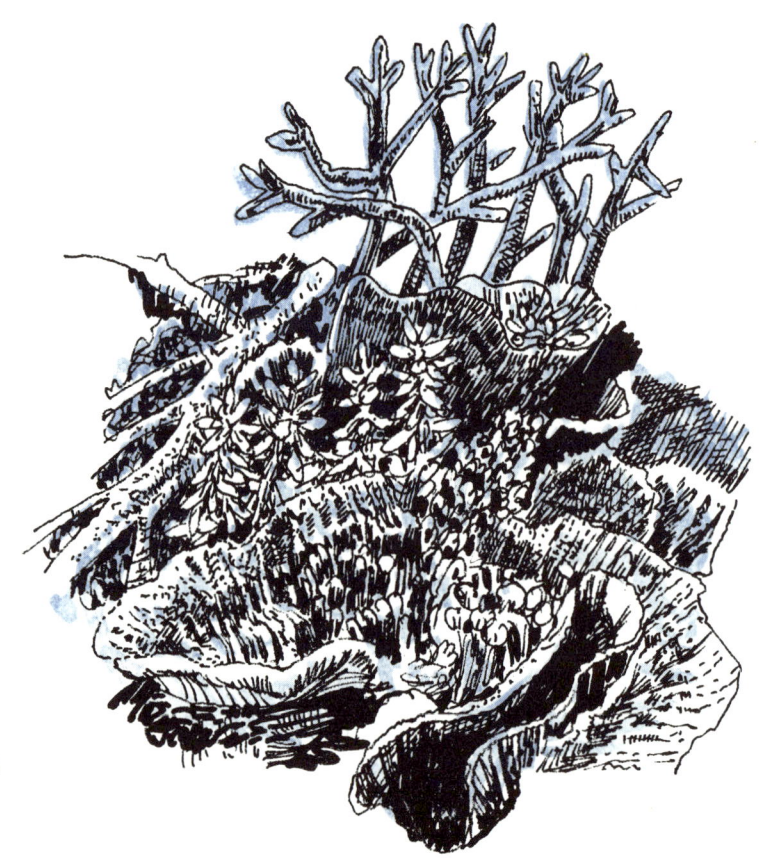

Anemones (flowers)

Sea Anemones

Flowers that shoot poisoned darts! That's what sea anemones do.

Sea anemones are just as colorful as the small flowers of that same name that grow in gardens and woods on land. Some are red, others white, and still others are many shades of lavender, green, and blue. However, anemones that grow in the oceans can be dangerous creatures.

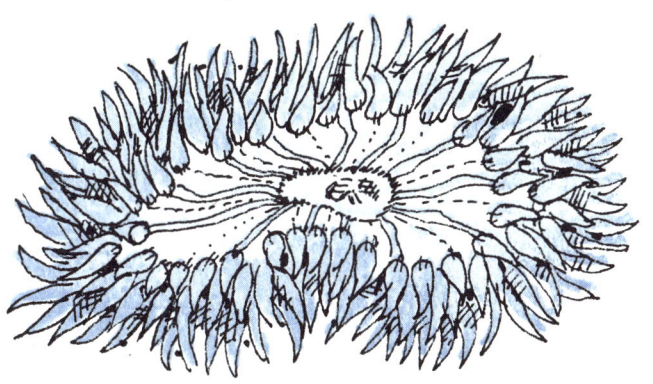

Sea anemones

Some sea anemones live in the deepest parts of the oceans. Others live close to the shores, clinging to rocks, shells, or anything else that might give them shelter.

Some of these pretty creatures live in small bunches. Others are spread like brilliant carpets on the Continental Shelf—the flat surface that stretches from the shores into the depths of the oceans. Still others look like delicate bouquets growing in cracks among water-covered rocks. Deep-sea divers say that some sunken ships are completely covered with these lovely sea flowers.

Although sea anemones look like flowers, they really are living animals. Some are as small as pansies; others are as large as sunflowers.

Their graceful "petals" sway in the currents of the water and lure small fish within range of the harmful darts that are shot at anything that swims by. These darts are tipped with a poison that stuns or kills the small fish. The petals then pull the fish into the middle of the flower. There, other petals force the fish into the sea anemone's mouth.

Jellyfish

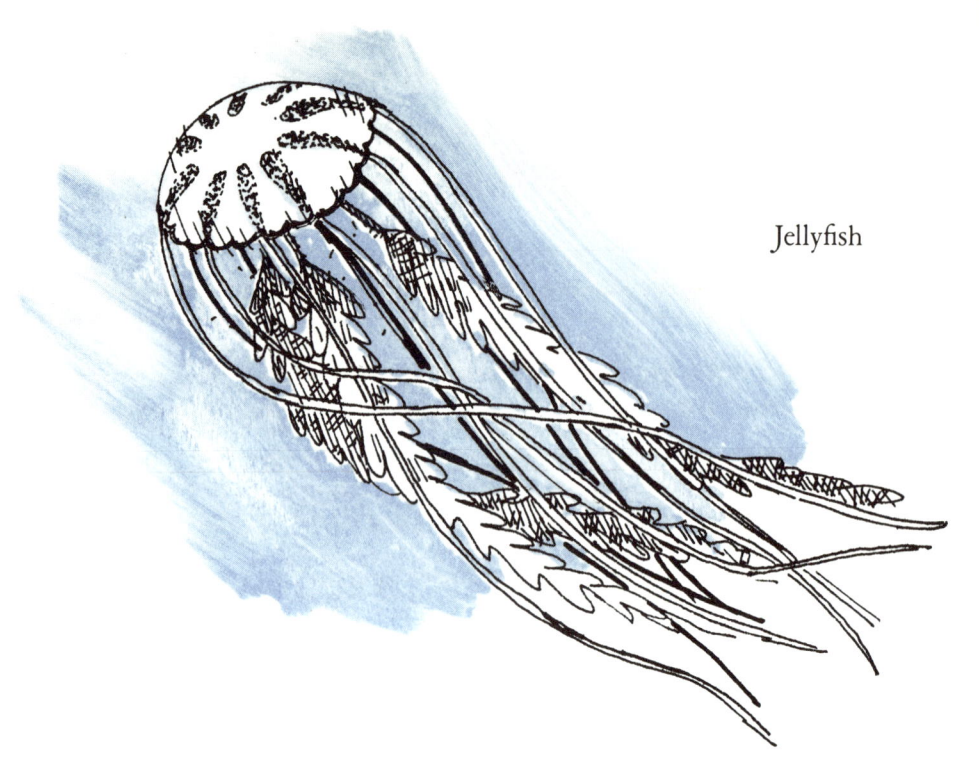

Sea anemones belong to the large jellyfish family. Each one is composed of a hole, or mouth, in the center that is surrounded by the petals. This hole leads to a stomach where the food is digested.

Although sea anemones are able to move slowly from place to place, most of them remain attached to one spot. In some areas along the shores and in deep waters there are so many anemones, of all kinds and colors, that they look like large flower gardens.

If a sea anemone is touched, the petals pull back into the center, and it looks like a dull, gray ball. As soon as the touch is taken away, the petals open again, and it becomes a lovely flower.

An amazing thing about sea anemones is that they have pets. Although their sharp darts are usually shot into anything that comes close to them, some fish can swim in and out among the dangerous petals without being harmed. These are the pets, and they are called damselfish.

Sea anemone and damselfish

Damselfish are small—only a couple of inches long—and are a bright orange color with white stripes across their bodies.

Perhaps darts are shot into the damselfish as they swim slowly through the sea-flower gardens, but they don't seem to be hurt by them.

Along the shores of Florida some anemones make pets of clown fish and shrimps. These small creatures also swim and crawl among the waving petals and don't seem to be hurt by them.

Clown fish

Although land anemones are grown from seed, like any other flower, the sea anemone is born. However, scientists don't know exactly how the sea anemone reproduces itself. It is now thought that some of the petals are male and others female. Several times a year a sea anemone drops small living organisms, called polyps. These cling to any surface they fall on, and those that aren't eaten by fish grow into pretty sea flowers.

Although sea anemones are dangerous when alive and in the sea, they soon lose their poison and die when brought into the air. When tides bring them ashore, they quickly dry up and look like ordinary seaweeds.

So when you are at the seashore, be careful when you wade in the water. What might look like a bouquet of pretty flowers could be little archers ready to sting you with their tiny arrows.

Their poison seldom kills anyone, but the stings do hurt, and many of them could make you very ill.

Octopus

The Octopus

Many people think of octopuses as huge creatures with waving long arms that drag ships into the depths of the sea. That isn't true. Some octopuses are rather large, but most of them are small. Small as they are, they can be dangerous.

The octopus has eight arms that seem to extend from the back of its head. But these arms are really attached to a very small body. This body is so small that, when the octopus swims or walks, it looks as if it had only the head and arms.

Near where the head seems to join the arms, the octopus has two eyes that look almost human. Below the eyes, and almost hidden in the arms, is the mouth. It is a very small opening, and looks like the beak of a parrot.

On each of the eight arms there are two rows of flat suckers. These are as sensitive as fingertips. When a fish comes within reach, the arms grab it, while the suckers hold it tightly until it dies. Then the arms take it to the parrot-mouth, which eats it.

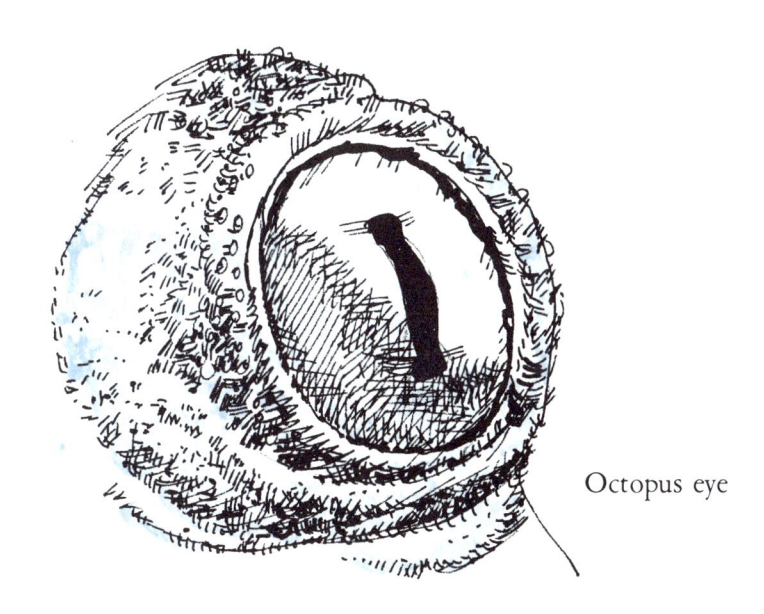

Octopus eye

Octopus eggs

On some octopuses there is a webbing between the arms. This is used in several ways. Onc way is to throw the web over a victim, like a net, and hold it until the arms and suckers can grab it.

Another way an octopus uses its web is to hide its eggs from an enemy. A female octopus lays her eggs in clusters, like bunches of grapes, usually on a rock or in a cave. She is a very good mother, and takes excellent care of her eggs. She washes them carefully and stands guard to keep them from danger.

When a mother octopus thinks an enemy is near, she spreads her arms so that the webbing covers the eggs. All an attacker sees is a tiny parrot-mouth ready to bite to protect her young ones.

Octopuses are shy, and hurry away when a swimmer comes near them. Sometimes they rise on their eight arms and scurry across the ocean floor. Usually they swim by shooting a stream of water from an opening in the backs of their heads. This propels them through the sea as though they were jet planes. They can also shoot a cloud of black material from their ink sacs, and disappear in a smoke-like screen.

Octopuses are of all sizes. Some are only a few inches long. Others will grow until they weigh a ton, and have arms ten feet long.

Octopuses are blamed for many horrible things that are really done by their cousins, the squids. Squids also have eight arms that are attached to tiny bodies. But two of their arms grow to great

Squid

lengths. Some that have been measured stretch for thirty feet.

A squid will attack anything it sees, even whales. Many people believe that the stories of sea serpents began when sailors in olden times saw a fight between a whale and a huge squid. It is probable that a whale might tear off one of the long arms of a squid, and it would look like a long neck swimming away.

Luckily giant squids live only in the deepest

parts of the oceans, and are rarely seen near the shores. Octopuses, however, live mostly in the shallow waters of all the seas, except those at the North and South Poles. They will not attack anyone, but they bite when they are disturbed.

Octopuses can change color, and become the same tint as the rocks and seaweed they hide in. They can change from dark brown to light blue and pink in a few seconds. That makes them hard to see, and so they are easily stepped on.

It is the mouth of an octopus that is dangerous. Its tiny jaw contains a poison that stuns fish and can cause pain and sickness to swimmers and waders.

The most dangerous of the octopuses is the smallest one. It is called the blue-ringed octopus because of the lovely blue color around the edge of its head. It is only two inches from the head to the end of its arms when it is fully grown. But if a wader or swimmer steps on one and receives a bite, death comes in about five minutes.

Plant plankton, much enlarged

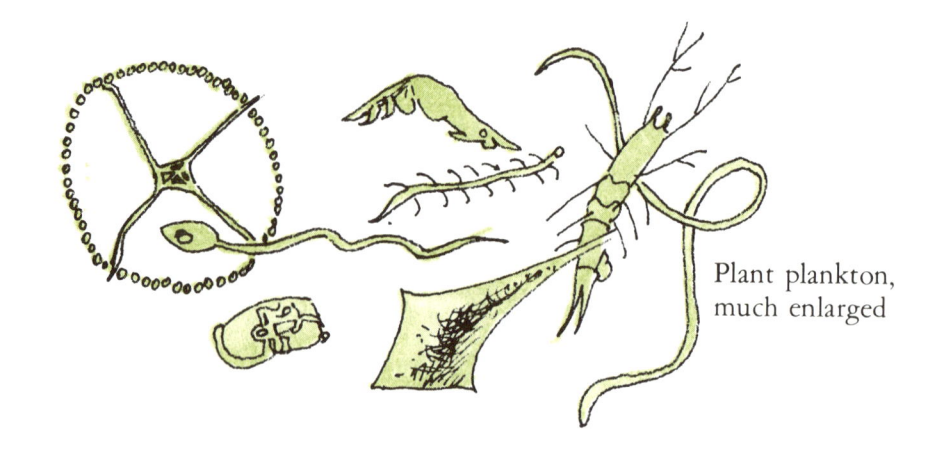

Corals

Along the southern coast of Florida, and other tropical shores, carpets that look like glittering jewels extend into the sea. These are corals of all colors.

Sometimes corals are mistaken for sea anemones. But there is a great difference between them. Anemones are like the plants that grow on land. And just as land plants live by feeding on the food in the earth, sea anemones eat plankton.

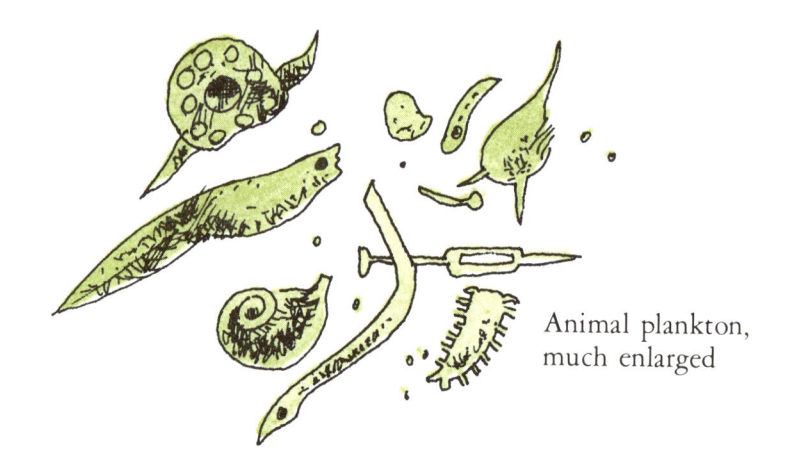

Animal plankton, much enlarged

Plankton is called the soup of the seas and oceans, because it is made up of countless numbers of tiny plants and animals mixed with the salt water. These living things are so small that they can be seen only through a microscope.

But corals are like land animals. When they are fully grown, most of their food is fish and even larger sea life.

Some corals sting the creatures that swim past them, stunning or killing them with the poison in their lovely flowers. Other corals catch the fish in long, leaf-like arms that hold them until they die.

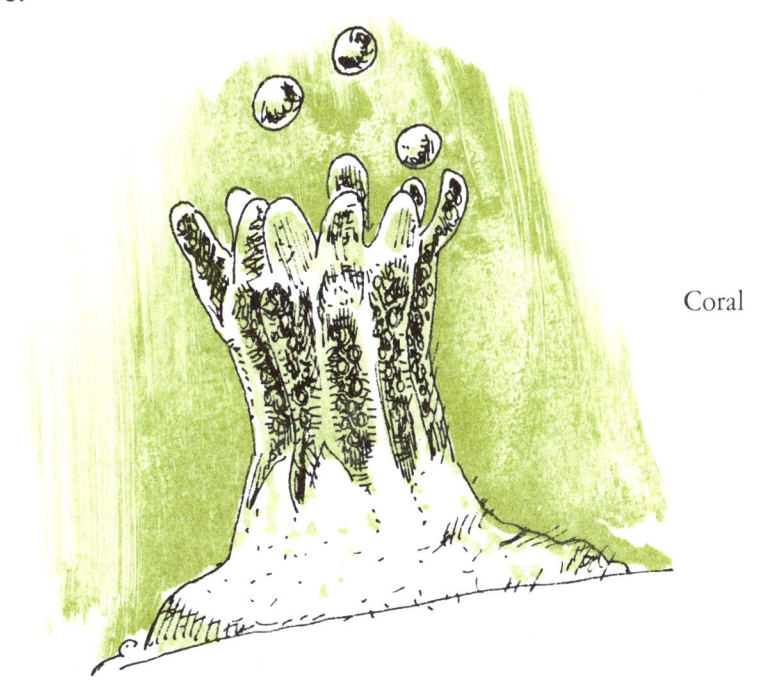

Coral

Corals can drift and swim, but most of them stay in one place all their lives. Corals build communities, like cities. When these coral cities join one another, they form the reefs that stretch into the ocean from tropical seacoasts.

One of the largest and most famous of coral reefs is the Great Barrier Reef that lies along the northeastern coast of Australia. This reef is more than a thousand miles long, and reaches into the ocean for hundreds of miles.

Great Barrier Reef

Coral reefs are a little like the rain forests of South America. They are just as thick, and just as full of life. Thousands of kinds of corals live in them, and these attract many thousands of different, strange, and beautifully colored fish.

Corals are born in a strange way. At various times of the year both male and female corals put cells, called seeds or buds, into the water. The cells drift with the currents and finally meet and mix. Then they settle onto the limestone base of the reef. This base is really made up of the skeletons of corals that have died. Corals live about a year; then their bones become part of the reef.

Most kinds of corals need sunshine in order to live, even though the sun's rays can reach them only through the water. That's why the most beautiful of them are found in the shallow water along tropical shores.

However, some other types of corals breed and live in the darkest, coldest, and deepest parts of

Coral community

the oceans. These corals need no sunlight, and are only rarely seen by divers who descend to great depths.

One of the amazing facts about corals is that mothers and daughters stay together to form one community, while fathers and sons form another. Each community puts cells into the water so they can meet and mix and produce new corals. But when these new corals reach a certain size, they separate and make more male and female communities.

As with many of the small dangerous creatures that live in the seas, corals are harmful only when they are disturbed. Most divers who work in and around coral reefs wear thick-soled shoes, so that if they accidentally tread on any of these beautiful sea jewels, they won't be stung with their poison.

But many corals do live very close to shore, and therefore are a danger to those who swim or wade in shallow water.

Be wise. Always wear some kind of foot covering when visiting the seashore, especially if you go into the water.

The Tilefish

The tilefish is found in a narrow band of water on the edge of the warm Gulf Stream, from Nova Scotia to the lower tip of Florida.

The strange thing about the tilefish is that since it was first seen, about a hundred years ago, it has appeared and disappeared from fishermen's nets a number of times.

For many years not one tilefish will be caught. Then they will appear again, and for a long time millions of pounds of them will be sold in the fish markets of the North.

Why does this strange fish appear and then disappear?

The people who study the creatures that live in the oceans believe they have the answer to the

Tilefish

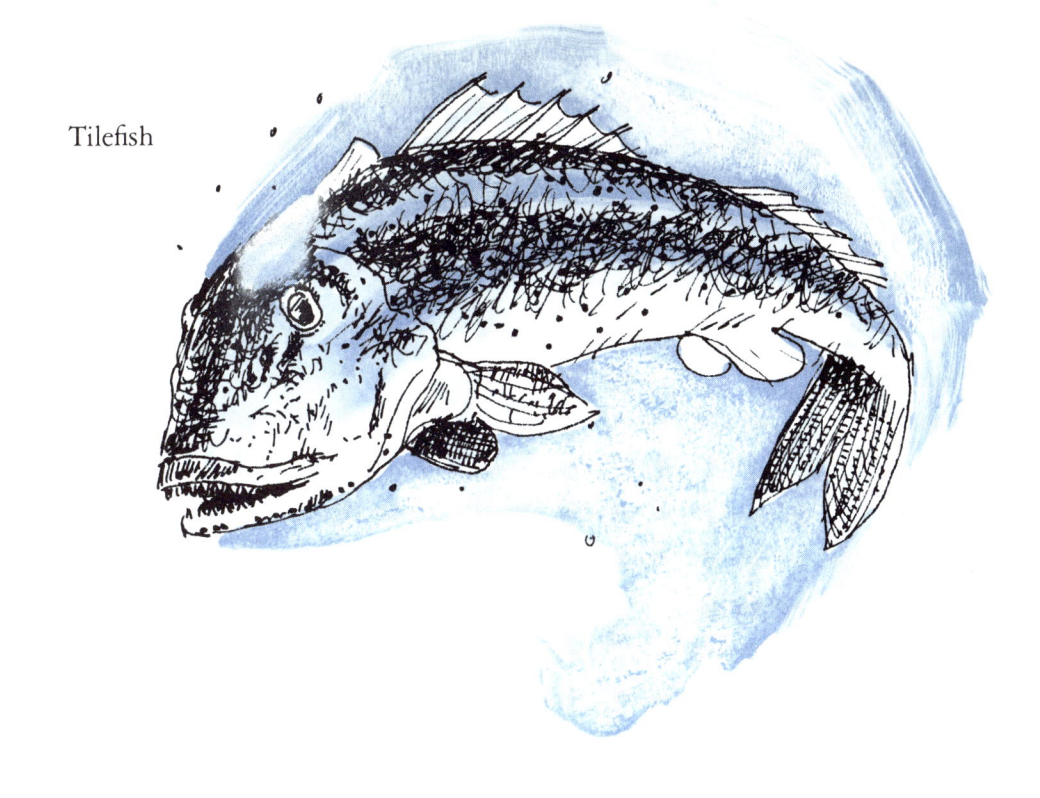

mystery. And the explanation is as strange as the fish themselves.

The story of the tilefish begins back in May, 1879, when Captain William H. Kerby found one strangely colored fish in the nets filled with codfish that his men pulled aboard. The fish was almost three feet long, and weighed 25 pounds. It was striped with bright colors of red, yellow, and blue, and stood out brightly from the gray

Head of tilefish, showing sharp teeth

codfish. It had a huge head, with a wide mouth that was lined with very sharp teeth in the shape of cones. Although Captain Kerby had been a fisherman all his life, he had never seen a fish like this one in northern waters.

Some strange fish are poisonous when eaten. But Captain Kerby was a brave and curious man. He had the ship's cook prepare the fish; then he and some of his crew ate it at dinner. It tasted delicious, and no one became ill.

Every day after that the nets brought in more and more of the odd fish. Soon they were being sold in New England fish markets as a delicacy.

Captain Kerby sent some of the fish to the Bureau of Fisheries. The scientists there said it was a new kind of fish that was related to a tropical type found only in the Gulf of Mexico. But they couldn't explain why it began to appear in the cold waters of the North. They gave it a long Latin name, which the fishermen turned into "tilefish."

Then, only three years later, on March 3, 1882, a Captain Lawrence suddenly found his fishing boat, *Plymouth,* sailing through a sea covered with dead tilefish. There were so many of them that the captain had difficulty ploughing his ship through them. The blanket of dead tilefish stretched across the water for 69 miles!

During the next ten years not one tilefish was caught.

Then, in 1892, eight tilefish were found in the nets. The next year 50 more were seen. Tilefish became more common during the next ten years, and millions of pounds were caught. Then they suddenly disappeared again! And again, almost ten years later, they appeared in large numbers!

Finally, after years of study, the official explanation of the mysterious appearances and disappearances of the tilefish is this: These fish like warm water. They also do their feeding on the sea bottoms, but only in water no more than 150 feet deep. When the Gulf Stream comes close to the shores, the tilefish are plentiful. But when, as has been happening every ten years or so, the warm waters move farther from shore, the tilefish die. They can't follow the warm water into the depths of the ocean and they quickly perish in cold water.

Now these strange fish are plentiful again, especially between Nantucket and Cape May, New

Jersey. But they might disappear at any time. It all depends on those uncertain currents of that narrow band of warm water that flows northward out of the Gulf of Mexico.

Although tilefish are lovely to look at, and delicious to eat, they are very dangerous to handle. Their large teeth can cause deep, painful wounds.

When dead tilefish are cast up on the shore, it's best to bury them in the sand. There they will become food for the millions of small sea creatures that live beneath our beaches.

Lamprey

The Lamprey

Its nose is on top, where its head should be. But it has no head! Its face is a flat space filled with dozens of small suckers, with a rough tongue in the center. It is called the lamprey. It is one of the most harmful of the small creatures that live in the seas, and isn't very pretty to look at.

The lamprey attacks other fish by attaching the flat surface of its "face" to their bodies. Then the powerful suckers drain the blood from the victim.

These harmful fish are found in all the oceans of the world. But they can also live in fresh-water rivers and lakes. There they cause great damage to the other fish that live in fresh water.

Lampreys are like salmon in one way. Every spring, full-grown lampreys enter rivers to lay their eggs. Many of the newly born lampreys return to the sea, then come back again to the same rivers a year later to lay their own eggs. But some of the lampreys remain in the rivers, where they attack trout and other fresh-water fish, and kill many of them.

Also, great numbers of lampreys make their way into lakes. There they kill almost all the fish that live in the fresh water.

Here is one example of the damage they can do: For many years, lampreys came up the St. Lawrence River to lay their eggs. Some of them entered Lake Ontario, one of the Great Lakes that is closest to the sea. However, they couldn't enter the other Great Lakes because Niagara Falls, at the end of Lake Ontario, was a barrier they couldn't cross.

But a number of years ago, the Welland Canal was built. It joined Lake Ontario to Lake Erie, and made it possible for large ships to pass around Niagara Falls and enter the other Great Lakes. The canal also allowed the lampreys to swim into those huge fresh-water lakes. The result was the death of many thousands of trout and other fish.

In 1939, commercial fishermen caught almost two million pounds of trout. In 1951, twelve years later, only 25 pounds were caught. The rest had been killed by the lampreys.

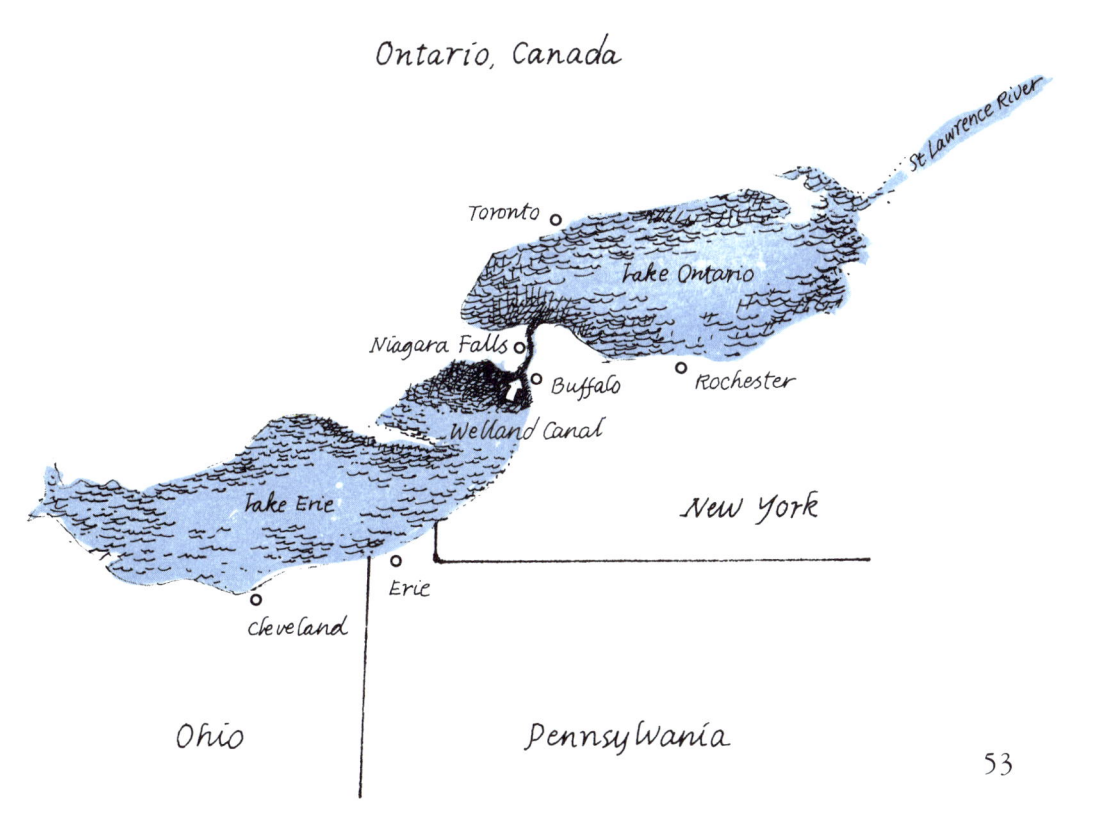

Since that time scientists have been experimenting with ways to stop the lampreys from entering the lakes from their spawning places in the rivers. They tried poison and chemicals. But nothing stopped the lampreys. Then, a few years ago, they found the solution to the problem. They built weirs, or screens, across the rivers. These weirs are made up of hundreds of electric wires that hang deep into the water. The wires set up an electric current that spreads across the rivers like an invisible curtain.

The electric screens have worked very well. Lampreys trying to pass through the invisible curtain are stunned and turned back, or killed, by electric shocks.

Year after year, fewer lampreys have managed to enter the Great Lakes. For some strange reason, lampreys never spawn in the lakes. They prefer the rivers. Slowly the lamprey population is dying away. Soon there will be no more of these harmful fish feeding on fresh-water creatures.

Lampreys grow to a length of one and a half to three feet. Each female lamprey lays several thousand eggs at a time.

In the past it was thought that lampreys were good to eat. But they can be poisonous to some people. In the year 1135, Henry I of England died from eating a lamprey. But they do have some good use. When they are caught they are ground up and used as fertilizer on our farms.

Lampreys swim very close to the shore when they are in rivers and lakes. Many waders and swimmers have been attacked by their suckers. Lampreys have no eyes. But when they bump against flesh of any kind the suckers begin their harmful work of draining the blood from their victims.

Lamprey eggs

Sea Urchins

Sea urchins look like pincushions. They are round, with many thin spikes sticking up from their tops. There are more than 800 different kinds. Some are as small as golf balls. Others are as large as grapefruit. They are of all the colors of the rainbow, depending on what parts of the oceans they live in.

All sea urchins are divided into five parts, like a sliced pie. Each part has a mouth on the underside of the body, or ball.

Some of the spikes on sea urchins are tipped

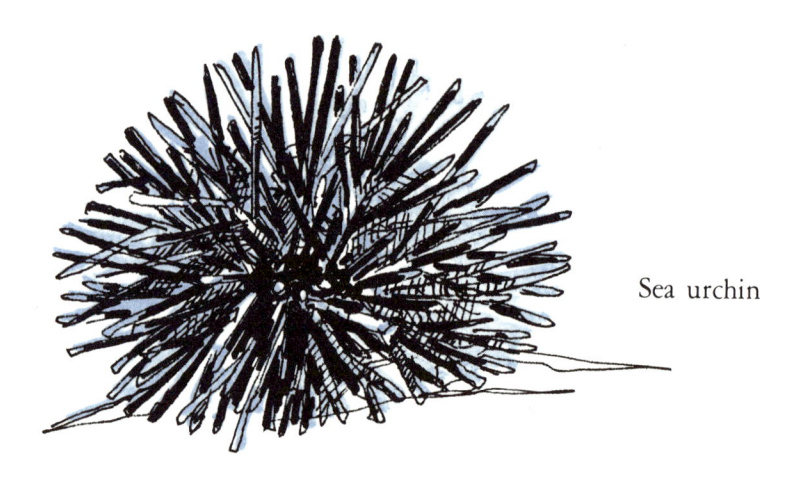

Sea urchin

Sea urchin

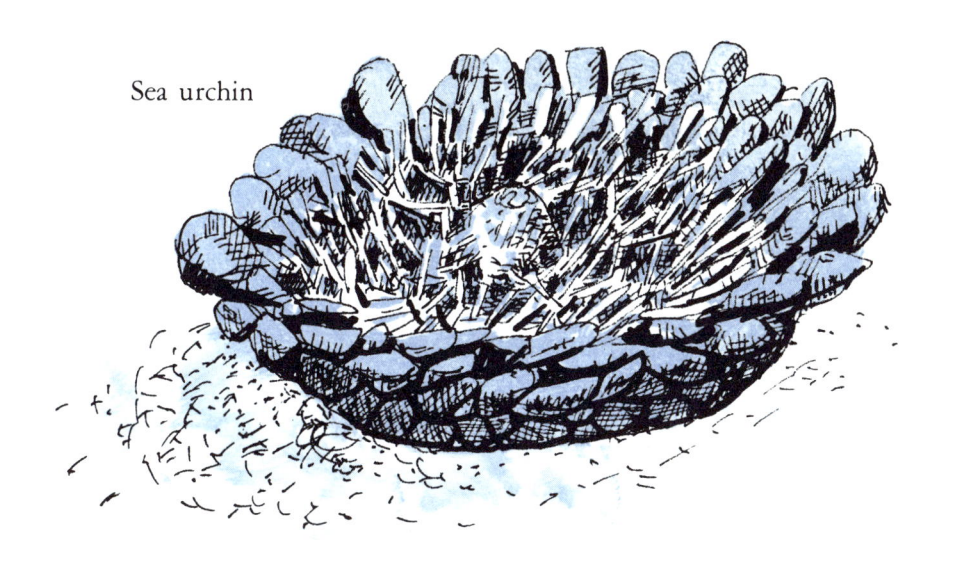

with a poison that kills the fish they feed on. Other spikes have small suckers on their ends. These push the food into the mouths. Still other spikes are like walking canes that help them move across the ocean floor.

Sea urchins are the most plentiful of all the sea animals. Half of the ocean floors are covered with these strange, spiky balls.

Sea urchins first appeared in the oceans more than 500 million years ago. In that faraway time, the seas covered almost all of the earth. That's why the bones of sea urchins are still found in the mountains and on the plains of almost all the countries of the world.

In past ages many people thought sea urchins were thunderbolts thrown down to the earth during storms. And because these ancient people believed lightning wouldn't strike twice in the same place, they saved the spiny balls as protection against storms. They even thought sea urchins would prevent milk from turning sour, so many dried sea urchins were kept in the dairies.

Sea urchin

Sea urchin

Some people were buried with sea urchins beside them, for they were thought to bring good luck in a future life.

A female sea urchin gives birth to millions of tiny eggs at one time. A great many of these eggs become food for fish of all sizes, even whales. But some of the eggs rest on the ocean floors and, in a few weeks, become full-grown sea urchins. Soon these place more millions of eggs into the water.

Although many sea creatures are dying because

of the pollution in the waters of the oceans and seas, sea urchins don't seem to mind foul water at all. They seem to thrive on the weeds and other plants that grow in it.

Sea urchins eat many fish, but they really like to graze on sea grasses, just as cows do on land grass. The sea urchins that live near the shores cling to the waving strands of seaweed and the leaves of ocean weeds.

Sea urchins are very plentiful along both coasts of the United States. When they are washed up

Sea urchin shell

Sea urchin shell

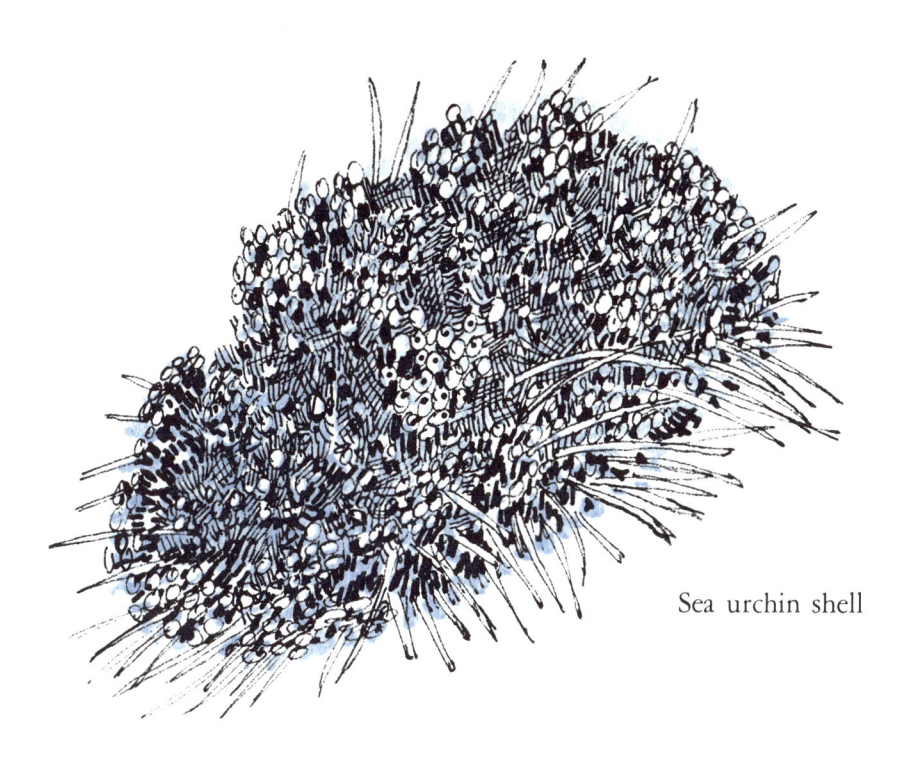

onto the shores, they quickly die. But even then they should be handled carefully, because their spikes carry poison for many days. The poison isn't strong enough to kill anyone who is struck by a spike, but it will cause pain and illness that lasts several days.

Perhaps some day a use will be found for these mysterious round balls that cover almost all of the sea floors. Until then it is enough that their many eggs are food for other creatures that live in the oceans.

There is a very good chance that you could go to the seashore many times to wade or swim without meeting any small deadly sea creatures. But you should be aware that they do exist.

Be careful, not only when you meet any of those described in this book, but also when you see any strange creatures in the water or lying on the beach. No matter how small or pretty they might seem to be, do not touch them or go close to them unless you know what they are.

Remember, no matter how harmful some sea creatures might be to people, they all serve an important purpose in the chain of life. Some of them are food for animals that are larger than they are, while they, in turn, eat creatures that are smaller.

And so the living things on our earth keep a balance, with each kind depending on another kind for survival. It is when we interrupt or disturb these creatures that they defend themselves, and are dangerous to us.

Index

anemones, 25, 31
arms, of starfish, 7, 10;
 of octopus, 32-35;
 of squids, 35-36;
 of corals, 39
Atlantic Ocean, starfish in, 12
Australia, reef on coast of, 40

barbs, of man-of-war, 15, 16;
 see also darts *and* spikes
barnacles, starfish and, 9
blue-ringed octopus, 37
Bureau of Fisheries, 47

camouflage, 19-20, 37
Cape May, N. J., tilefish and, 48-49
clown fish, 29
colors, of starfish, 11;
 of scorpion fish, 19;
 of sea anemones, 25, 28;
 of octopus, 37;
 of corals, 38;
 of tilefish, 45;
 of sea urchins, 56
communities, of corals, 40, 42
Continental Shelf, sea anemones on, 26
corals, 19, 24, 38-43

damselfish, and sea anemones, 29-30
danger, from sea creatures, 5, 62;
 from starfish, 12;
 from man-of-war, 13, 16-17;
 from scorpion fish, 20-21;
 from sea anemones, 31;
 from octopus, 37;
 from tilefish, 49;
 from lamprey, 50
darts, of sea anemones, 25, 27, 29-31;
 see also barbs *and* spikes
dorsal fin, of scorpion fish, 21-24

eggs, of man-of-war, 14;
 of octopus, 34, 35;
 of lampreys, 52, 55;
 of sea urchins, 59, 61
eyes, of starfish, 7;
 of octopus, 33;
 lampreys and, 55

face, of lamprey, 50

females, of starfish, 10;
 of sea anemones, 31;
 of octopus, 34, 35;
 of corals, 41, 42;
 of lampreys, 55;
 of sea urchins, 59
fertilizer, lampreys as, 55
fins, of scorpion fish, 20, 21
fishermen, 7, 44-48, 53
Florida, sea creatures in waters of, 12, 16, 18, 30, 38, 44
food, sea animals and, 5;
 starfish and, 7-9;
 for humans, 9, 46, 47, 49;
 man-of-war and, 14, 15;
 and scorpion fish, 20-21;
 and sea anemones, 27-28, 38;
 of octopus, 33;
 for corals, 39;
 tilefish as, 46, 47, 49;
 for sea urchins, 57, 60;
 sea urchin eggs as, 59, 61

Great Barrier Reef, 40
Great Lakes, 52-54
Gulf of Mexico, sea creatures in 12, 18, 47, 49
Gulf Stream, tilefish and, 44, 48

habitat, of starfish, 12;
 of men-of-war, 16;
 of scorpion fish, 18;
 of sea anemones, 26;
 of squid, 36-37;
 of octopus, 37;
 of corals, 38, 41-42;
 of tilefish, 44, 48;
 of lampreys, 51, 52;
 of sea urchins, 57, 60
head, of octopus, 32-33;
 of tilefish, 46
Henry I, of England, 55
humans, starfish and, 9, 12;
 men-of-war and, 15-17;
 scorpion fish and, 21, 24;
 and sea anemones, 31;
 and octopus, 37;
 and corals, 43;
 and tilefish, 49;
 and lampreys, 55;
 and sea urchins, 61

ink sacs, of octopus, 35

jellyfish, 13, 28

Kerby, Captain William H., 45-47

Lake Erie, 52
Lake Ontario, lampreys and, 52
lampreys, 50-55;
   and Welland Canal, 52-54;
   death of, 54
Lawrence, Captain, 47
life, chain of, 6, 62

Maine, starfish and, 12
males, of sea anemones, 31;
   of corals, 41, 42
mouth, of starfish, 8;
   of scorpion fish, 23;
   of sea anemone, 27, 28;
   of octopus, 33, 35, 37;
   of tilefish, 46;
   of sea urchins, 56, 57

Nantucket, and tilefish, 48
New England, man-of-war and, 16
Niagara Falls, 52, 53
nomeus, 15
North America, shores off, 6
nose, of lamprey, 50
Nova Scotia, tilefish and, 44

octopus, 32-37
oyster borer, starfish and, 9
oysters, and starfish, 7, 9

pelicans, and starfish, 10-11
"petals," of sea anemones, 27-31
pets, of sea anemones, 29-30
plankton, 38-39
poison, 5;
   from starfish, 8, 12;
   from man-of-war, 15;
   from scorpion fish, 18, 21, 23;
   from sea anemones, 25, 27, 31;
   from octopus, 37;
   from corals, 39, 43;
   from lampreys, 55;
   from sea urchins, 57, 61
pollution, sea creatures and, 59-60
polyps, of sea anemones, 31
Portuguese man-of-war, 13-17

reefs, of corals, 40-41
reproduction, starfish and, 10;
   sea anemones and, 31;
   corals and, 41

sail, of man-of-war, 13-14, 16
St. Lawrence River, lampreys and, 52
salmon, lampreys and, 52
scientists, 9, 31, 47, 54
scorpion, 18
scorpion fish, 18-24
sea anemones, 25-31;
   and corals, 38
sea serpents, 36
sea urchins, 56-61
seagulls, and starfish, 10-11
shellfish, starfish and, 7, 9
shrimps, 7, 9, 30
size, of starfish, 10;
   of sea anemones, 26;
   of octopus, 32, 35, 37;
   of tilefish, 45;
   of lampreys, 55;
   of sea urchins, 56
spikes of starfish, 7, 11, 12;
   of scorpion fish, 19-21, 23;
   of sea urchins, 56-57, 61;
   see also barbs *and* darts
squids, 35-37
starfish, 7-12
stings, 5;
   of sea anemones, 31;
   of corals, 39, 43
stomach, of starfish, 8;
   of sea anemones, 28
streamers, of man-of-war, 13, 15-17
suckers, of octopus, 33, 34;
   of lamprey, 50, 55;
   of sea urchins, 57
superstitions, about sea urchins, 58-59

teeth, of tilefish, 46, 49
tilefish, 44-49
tongue, of lamprey, 50
trout, and lampreys, 53

webbing, of octopus, 34, 35
Welland Canal, 52-53
whales, and squids, 36